A LOOK AT PHYSICS

FORCES AND MOTION

BY KATHLEEN CONNORS

Gareth Stevens
PUBLISHING

CRASH COURSE

Please visit our website, www.garethstevens.com. For a free color catalog of all our high-quality books, call toll free 1-800-542-2595 or fax 1-877-542-2596.

Cataloging-in-Publication Data

Names: Connors, Kathleen.
Title: Forces and motion / Kathleen Connors.
Description: New York : Gareth Stevens Publishing, 2019. | Series: A look at physical science | Includes index.
Identifiers: LCCN ISBN 9781538221457 (pbk.) | ISBN 9781538221433 (library bound) | ISBN 9781538221464 (6 pack)
Subjects: LCSH: Motion--Juvenile literature. | Force and energy--Juvenile literature.
Classification: LCC QC127.4 C66 2019 | DDC 531'.6--dc23

First Edition

Published in 2019 by
Gareth Stevens Publishing
111 East 14th Street, Suite 349
New York, NY 10003

Copyright © 2019 Gareth Stevens Publishing

Designer: Samantha DeMartin
Editor: Kristen Nelson

Photo credits: Series art Creative Mood/Shutterstock.com; cover, p. 1 Jamie Wilson/Shutterstock.com; p. 5 (golf ball) Mikael Damkier/Shutterstock.com; p. 5 (scooters) Dmytro Zinkevych/Shutterstock.com; p. 5 (apple) Helioscribe/Shutterstock.com; p. 7 sirtravelalot/Shutterstock.com; p. 9 Vadim Sadovski/Shutterstock.com; p. 11 Monika Wisniewska/Shutterstock.com; p. 13 (inset) Cipariss/Shutterstock.com; p. 13 (main) Scimat Scimat/Science Source/Getty Images; p. 15 Ollyy/Shutterstock.com; p. 17 Everett Historical/Shutterstock.com; p. 19 Nazeri Mamat/Shutterstock.com; pp. 21, 23 Jim Lambert/Shutterstock.com; p. 25 De Repente/Shutterstock.com; p. 27 tammykayphoto/Shutterstock.com; p. 29 Mamontova Yulia/Shutterstock.com; p. 30 (rocks) Anatoli Styf/Shutterstock.com; p. 30 (baseball) Kris Petkong/Shutterstock.com; p. 30 (lawn mower) JNP/Shutterstock.com.

All rights reserved. No part of this book may be reproduced in any form without permission in writing from the publisher, except by a reviewer.

Printed in the United States of America

CPSIA compliance information: Batch #CS18GS: For further information contact Gareth Stevens, New York, New York at 1-800-542-2595.

CONTENTS

Push and Pull	4
Field Forces	6
Contact Forces	10
It's the Law!	16
The First Law	18
Unbalanced Forces	20
The Second Law	22
The Third Law	24
Measuring Force	26
Keep Going!	28
The Basics of Newton's Laws of Motion	30
Glossary	31
For More Information	32
Index	32

Words in the glossary appear in **bold** type the first time they are used in the text.

PUSH AND PULL

Anytime something is in motion, it happens because of forces. Forces are the pushes and pulls on an object. Sometimes these forces are balanced, or equal, and an object doesn't move. Other times, they're unbalanced, and balls roll or scooters go!

> **MAKE THE GRADE**
> As used in science, "motion" means "the change in position of an object."

FIELD FORCES

There are many different forces around us all the time. Field forces are one main group of these. These forces don't touch an object, but act on it at a distance. These forces include magnetic, electric, and **gravitational** forces.

> **MAKE THE GRADE**
> The **mass** of an object and how close the object is to the force are part of what **determines** the size of the force.

7

Gravity is the force that pulls objects down toward Earth's center. However, there's a small gravitational pull between all objects that pulls them toward one another. When one object has a large mass, the pull can be very strong.

> **MAKE THE GRADE**
> Electric and magnetic forces may **attract** or **repel** each other. Gravitational forces only attract.

CONTACT FORCES

Contact forces are another group of forces. These happen when two objects are touching. Friction is a contact force that stops motion. The tiny peaks and dips on an object's surface catch on the peaks and dips of other surfaces.

MAKE THE GRADE
The heavier an object, the more friction there will be between it and other objects.

Objects with **rough** surfaces will have more friction. But friction will even occur between objects that appear to have smooth surfaces! That's because if you look closely enough, no surface is entirely smooth.

MAKE THE GRADE
When surfaces rub together, friction between them causes heat. Friction makes the **molecules** on the surfaces speed up and have more energy!

CLOSE-UP OF A PAPER TOWEL

13

Applied force is a kind of contact force. This is any force that's put on an object, including by people. The kick to a ball is an applied force. So is the wind blowing through a window and knocking a glass over.

MAKE THE GRADE
More than one force is acting on all objects all the time!

15

IT'S THE LAW!

When the motion of an object changes, it changes because of force. Forces can start an object moving or stop an object from moving. There are laws that are always true about force and motion.

MAKE THE GRADE
Isaac Newton was the scientist who famously studied gravity. He explained some laws of motion, and they're named for him.

THE FIRST LAW

Newton's first law of motion is often called the law of inertia. Inertia is the **property** that makes matter **resist** changes in motion. It states an object at rest will stay at rest unless a force acts on it.

> **MAKE THE GRADE**
> The law also says an object that's moving will keep moving unless a force acts on it.

19

UNBALANCED FORCES

Newton's first law of motion has to do with objects that are being acted on by balanced forces. The forces are equal to one another. Newton's second law explains what happens to objects being acted on by unbalanced forces.

MAKE THE GRADE
An unbalanced force is one that is unequal to other forces acting on an object.

THE SECOND LAW

The second law states that unbalanced forces cause a change in motion, or acceleration, to occur. Acceleration is a change in an object's speed, direction, or both. How much an object will accelerate depends on the net force on the object and the object's mass.

MAKE THE GRADE
The net force is the total of all the forces acting upon an object.

THE THIRD LAW

Newton's third law of motion is about objects **interacting** with one another. Objects **exert** forces on each other, called action and reaction forces. When one object exerts force on another, the second object exerts an equal force in the opposite direction.

MAKE THE GRADE

Think of Newton's third law in an example. You exert a force on a chair when you sit in it. The chair also exerts a force on you as you sit. This allows you to stay in the chair!

MEASURING FORCE

The unit used to measure force is called the Newton (N), again named for Isaac Newton. Newtons measure both the size and the direction of a force, such as how hard you throw a ball and where you throw it.

> **MAKE THE GRADE**
> The force of an apple in your hand is equal to about 1 Newton!

KEEP GOING!

Once an object is moving, it's hard to stop it! This is a property called momentum. The more mass or speed an object has, the more momentum it has. The more momentum an object has, the harder it is to stop!

MAKE THE GRADE
"Frame of reference" has to do with something that isn't moving with respect to a person that can be used to figure out if something else is moving.

THE BASICS OF NEWTON'S LAWS OF MOTION

1ST LAW
Objects resist a change in motion.

2ND LAW
Unbalanced forces cause objects to accelerate.

3RD LAW
Two objects exert equal but opposite forces on each other.

GLOSSARY

attract: to draw nearer

determine: to be the cause for something

exert: to cause to have an effect or be felt

gravitational: having to do with gravity, or the force that pulls objects toward the center of Earth or attracts objects to one another

interact: to act on each other

mass: the amount of matter in an object

molecule: a very small piece of matter

property: a quality or feature of something

repel: to force away

resist: to stop or fight against

rough: not smooth or soft

FOR MORE INFORMATION

BOOKS

Ives, Rob. *Fun Experiments with Forces and Motion: Hovercrafts, Rockets, and More.* Minneapolis, MN: Hungry Tomato, 2018.

Kenney, Karen Latchana. *Forces and Motion Investigations.* Minneapolis, MN: Lerner Publications, 2018.

WEBSITES

Force and Motion: The Show
idahoptv.org/sciencetrek/topics/force_and_motion/index.cfm
Find out more about force and motion here.

Force & Motion
studyjams.scholastic.com/studyjams/jams/science/forces-and-motion/force-and-motion.htm
Watch a short video about force!

Publisher's note to educators and parents: Our editors have carefully reviewed these websites to ensure that they are suitable for students. Many websites change frequently, however, and we cannot guarantee that a site's future contents will continue to meet our high standards of quality and educational value. Be advised that students should be closely supervised whenever they access the internet.

INDEX

applied force 14
balanced forces 4, 20, 21
contact forces 10, 14
electric forces 6, 8
field forces 6
friction 10, 12
gravitational force 6, 8, 9
gravity 8, 16
inertia 18
magnetic forces 6, 8
momentum 28
net force 22, 23
Newton, Isaac 16, 17, 26
Newton's laws of motion 16, 18, 20, 22, 24, 25, 30
unbalanced forces 4, 20, 22, 30